Where I Live

My Planet

by Meg Gaertner

FOCUS READERS

SCOUT

www.focusreaders.com

Focus Readers is distributed by North Star Editions:
sales@northstareditions.com | 888-417-0195

Produced for Focus Readers by Red Line Editorial.

Photographs ©: JPL/NASA, cover, 1; Goddard/GSFC/NASA, 4 (top), 7, 16 (bottom left), 16 (bottom right); Shutterstock Images, 4 (bottom), 9, 11, 16 (top left), 16 (top right); iStockphoto, 13, 15 (top), 15 (bottom)

Library of Congress Cataloging-in-Publication Data
Names: Gaertner, Meg, author.
Title: My planet / by Meg Gaertner.
Description: Lake Elmo, MN : Focus Readers, [2021] | Series: Where I live | Includes index. | Audience: Grades K–1
Identifiers: LCCN 2019054514 (print) | LCCN 2019054515 (ebook) | ISBN 9781644933398 (hardcover) | ISBN 9781644934159 (paperback) | ISBN 9781644935675 (pdf) | ISBN 9781644934913 (ebook)
Subjects: LCSH: Earth (Planet)--Juvenile literature. | Planets--Juvenile literature.
Classification: LCC QB631.4 .G33 2021 (print) | LCC QB631.4 (ebook) | DDC 550--dc23
LC record available at https://lccn.loc.gov/2019054514
LC ebook record available at https://lccn.loc.gov/2019054515

Printed in the United States of America
Mankato, MN
082020

About the Author

Meg Gaertner is a children's book editor and author. She lives in Minneapolis, Minnesota. When not writing, she is usually dancing or spending time outside.

Table of Contents

sun

Earth

Earth

There are eight planets.

They move around the **sun**.

My planet is **Earth**.

Earth has a **moon**.

It moves around Earth.

I see the moon in the sky.

I see the moon at night.

moon

Parts of Earth

Some of Earth is land.

People live on the land.

land

Some of Earth is **oceans**.

They are large areas of water.

ocean

Earth has a layer of air.

The air goes around Earth.

We cannot see the air.

But we breathe it.

Home

Many people live on Earth.

We all live in different ways.

But Earth is our home.

Glossary

Earth

oceans

moon

sun

Index